SKATES ar

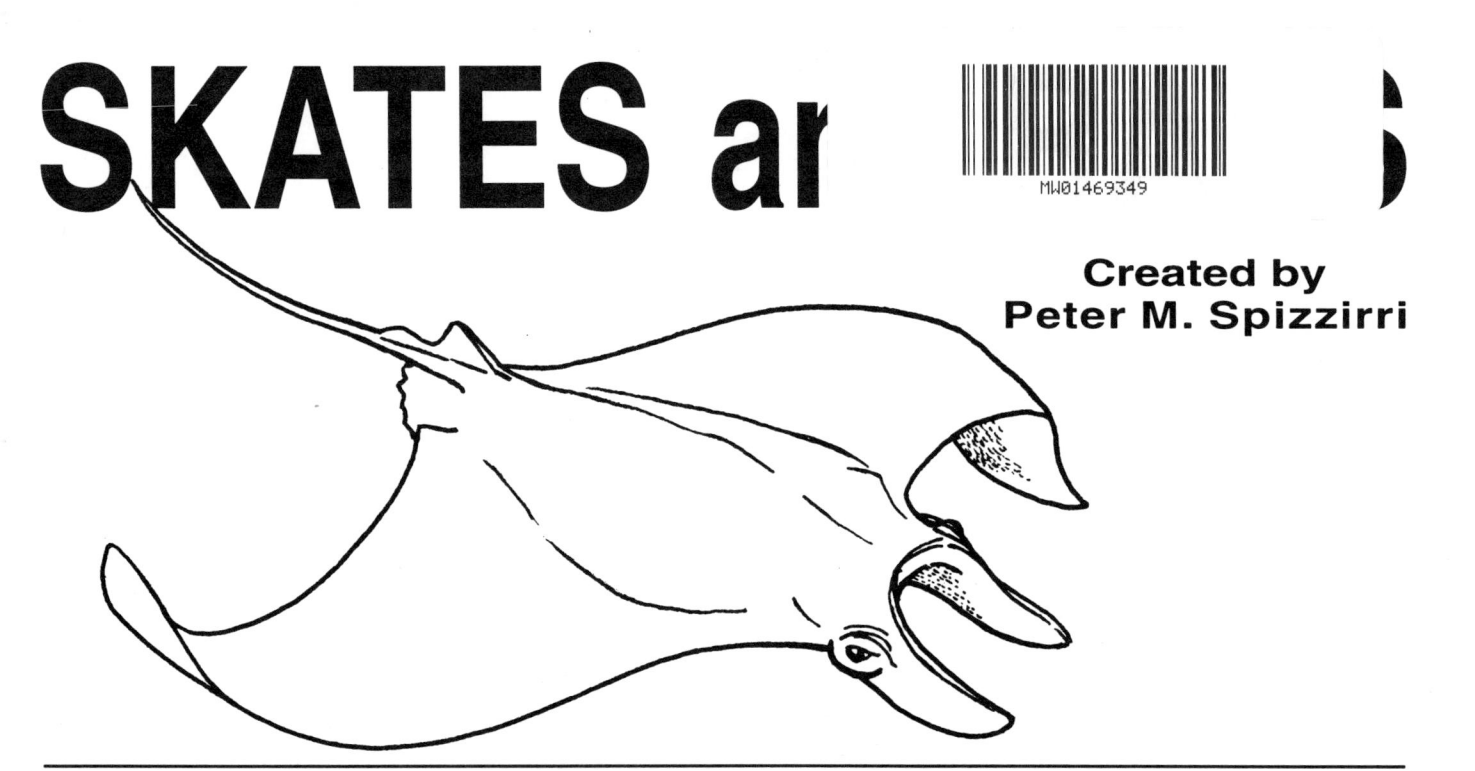

Created by Peter M. Spizzirri

Copyright, ©, 1996 by Spizzirri Publishing, Inc., All rights reserved.

This Is An Educational FACTS AND FUN Book Of SKATES & RAYS • Published by SPIZZIRRI PUBLISHING, INC., P.O. BOX 9397, RAPID CITY, SOUTH DAKOTA 57709. No part of this publication may be reproduced, stored in a retrievable system, or transmitted in any form without the express written consent of the publisher. All national and international rights reserved on the entire contents of this publication.
Printed in U.S.A.

PACIFIC ELECTRIC RAY
Torpedo californica

This Pacific ray has a flat disc shaped body with a 4 1/2 foot wingspan. It lives in shallow water and among the kelp beds. The halibut and herring it feeds on are stunned by its powerful electric charge. These rays have shown some aggressive actions toward divers, and may be dangerous.

••

The Pacific electric ray is found along the Pacific shore from Canada, South, all the way to Baja California.

PACIFIC ELECTRIC RAY
Torpedo californica

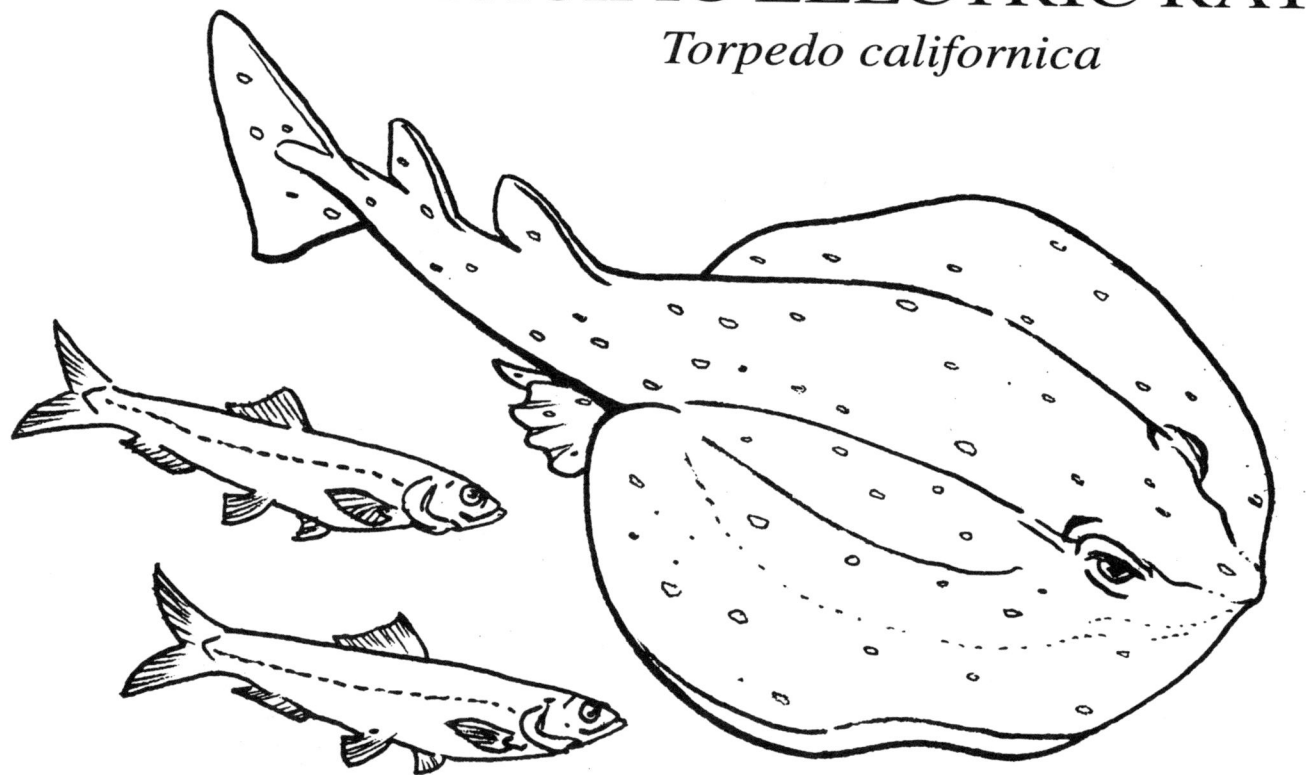

3

SPINY BUTTERFLY RAY
Gymnura altavela

Spiny butterfly rays are recognized by their huge 7 foot wingspan. It's short tail makes it look like a flying wing gliding through the water. These rays can change their color from light to dark, helping them blend into the background. They are more active than other rays, spending less time resting on the bottom.

••

The spiny butterfly ray is found in sandy shallow waters from Massachusetts, South to Argentina.

SPINY BUTTERFLY RAY
Gymnura altavela

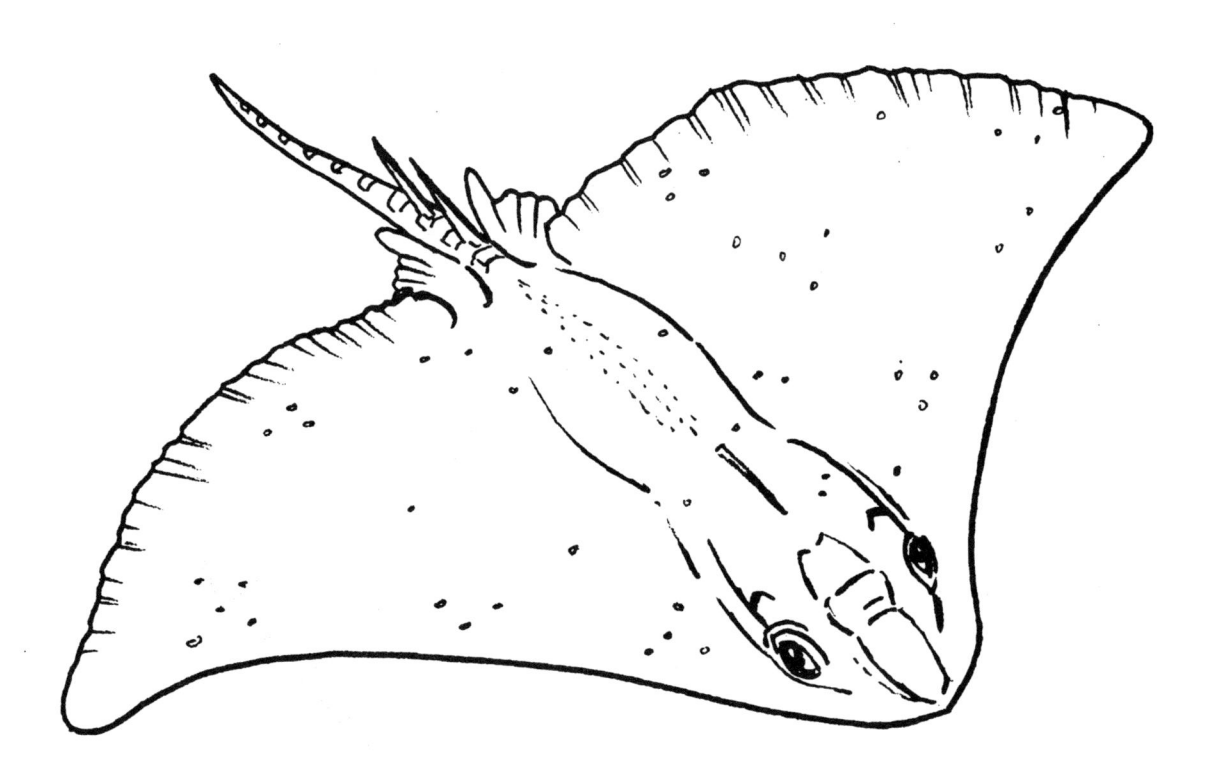

BULLNOSE RAY
Myliobatis freminvillei

The bullnose ray spends most of its time swimming in the open sea. It uses its three foot wingspan to stir up the ocean bottom, exposing the shellfish it hunts for dinner. The small poisonous spines, at the base of its tail, are not very useful as defense weapons.

• •

The Bullnose ray is found in the Gulf of Mexico, the Caribbean and the Atlantic Ocean from Boston to Brazil.

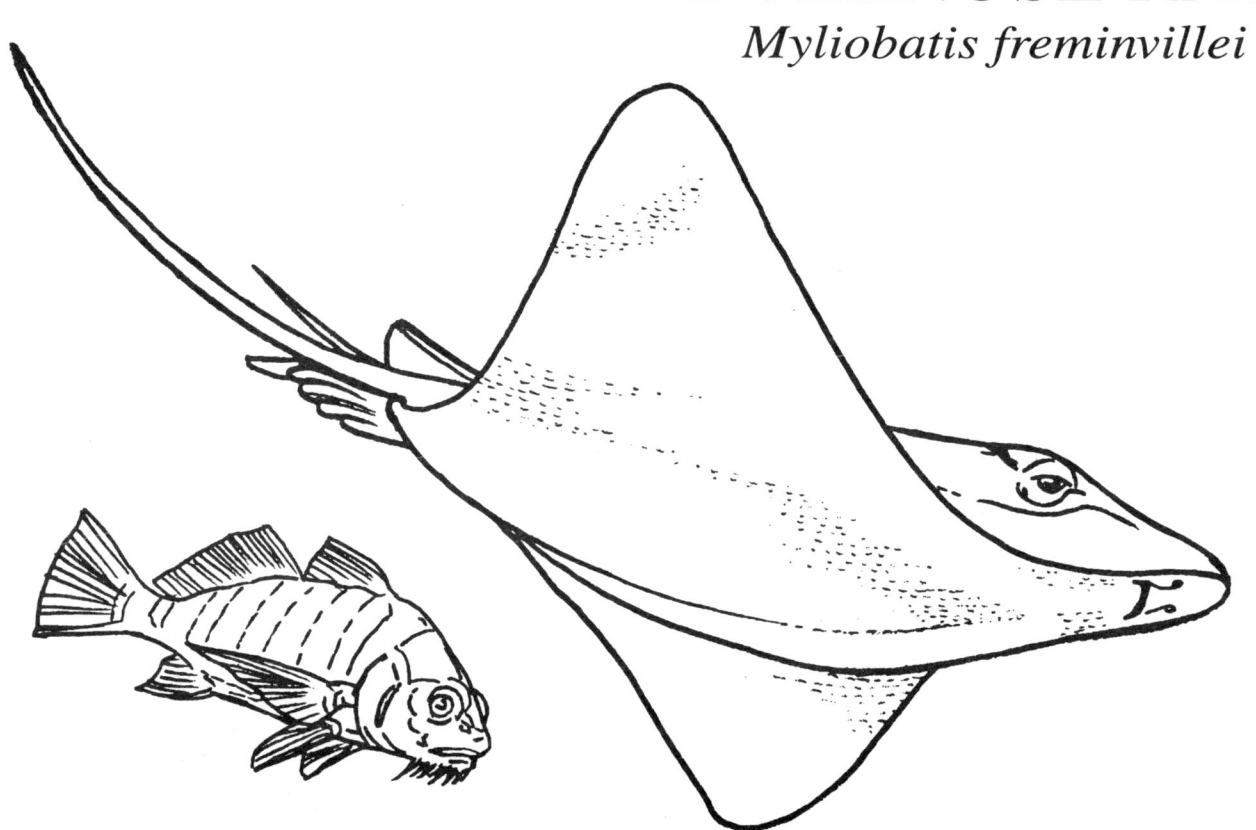

BULLNOSE RAY
Myliobatis freminvillei

7

SOUTHERN STINGRAY
Dasyatis americana

The five foot long southern stingray can be dangerous. They partially bury themselves in the sand close to shore. If bothered or stepped on, they inflict a serious wound with their poisonous tail spine. The poisonous spine points upward at the base of its whip-like tail.

••

The southern stingray lives in the Gulf of Mexico, the Caribbean and the Atlantic Ocean from New Jersey to Brazil.

SOUTHERN STINGRAY
Dasyatis americana

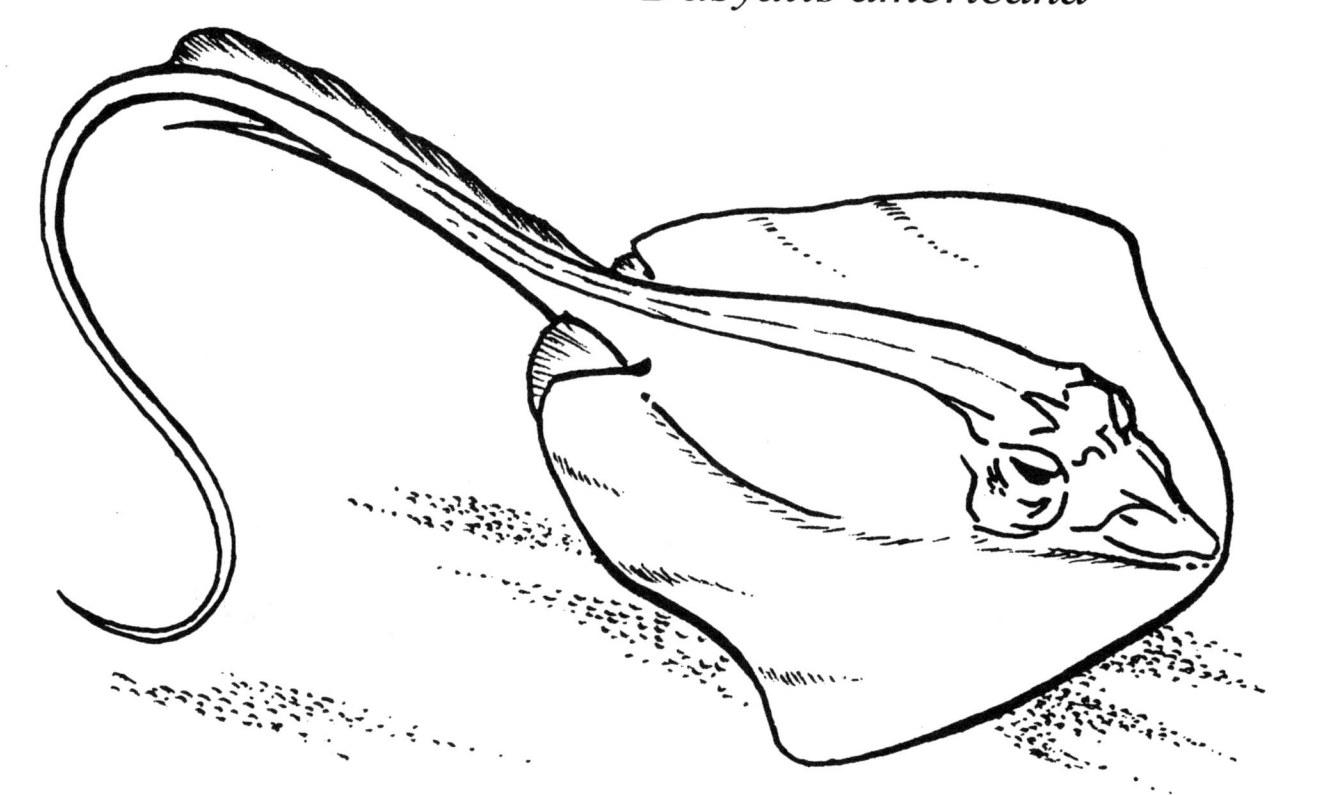

LESSER ELECTRIC RAY
Narcine brasiliensis

An overall pattern of large spots covers the lesser electric rays 18 inch body. This ray can give an electric shock of 37 volts, which is only used to defend itself. It feeds on crustaceans, worms and small animals that it catches without needing to use its electric shock.

••

This electric ray is found in the Gulf of Mexico, the Caribbean and the Atlantic Ocean from the Carolinas to Argentina.

LESSER ELECTRIC RAY
Narcine brasiliensis

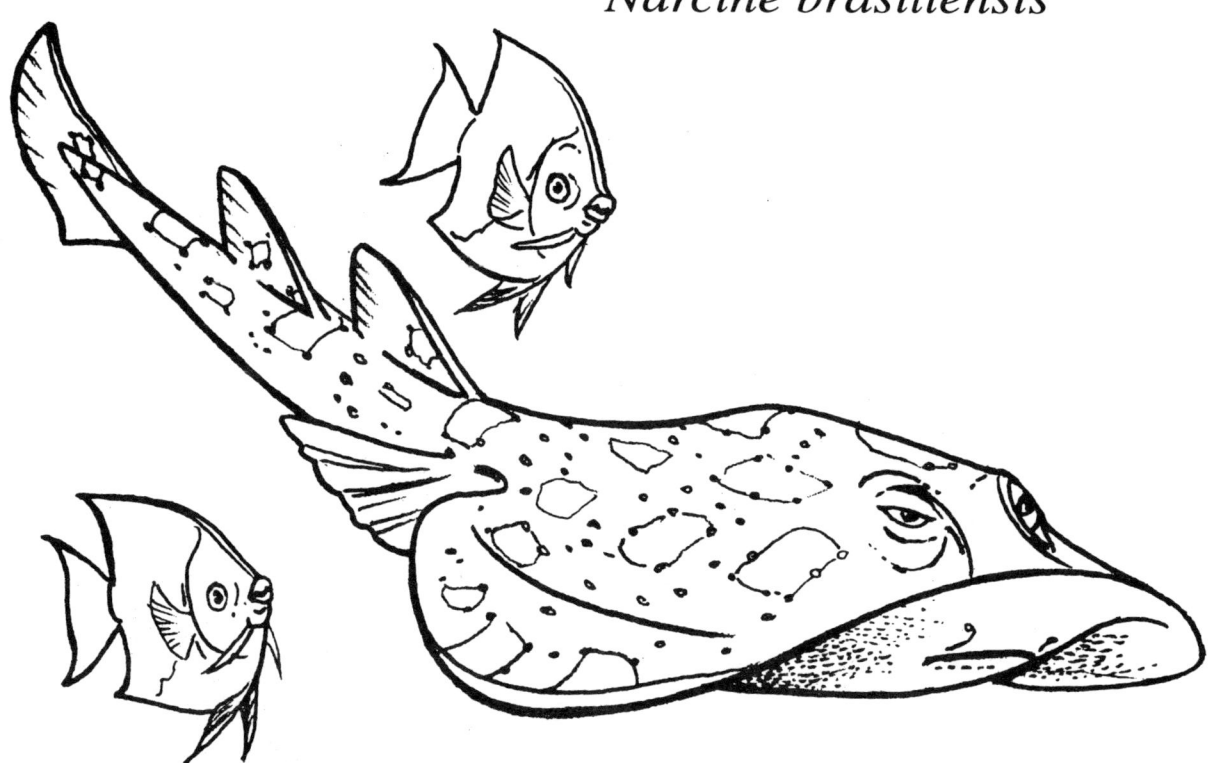

11

ATLANTIC TORPEDO RAY
Torpedo nobiliana

The large Atlantic torpedo ray can be six feet long and weigh 200 pounds. They spend most of their time on the ocean bottom because they are not good swimmers. This ray has a special muscle near its head that produces a powerful electric charge of 220 volts. Ouch!

••

The Atlantic torpedo ray is found on the Atlantic Coast from Nova Scotia, South to Florida and in the Gulf of Mexico.

ATLANTIC TORPEDO RAY
Torpedo nobiliana

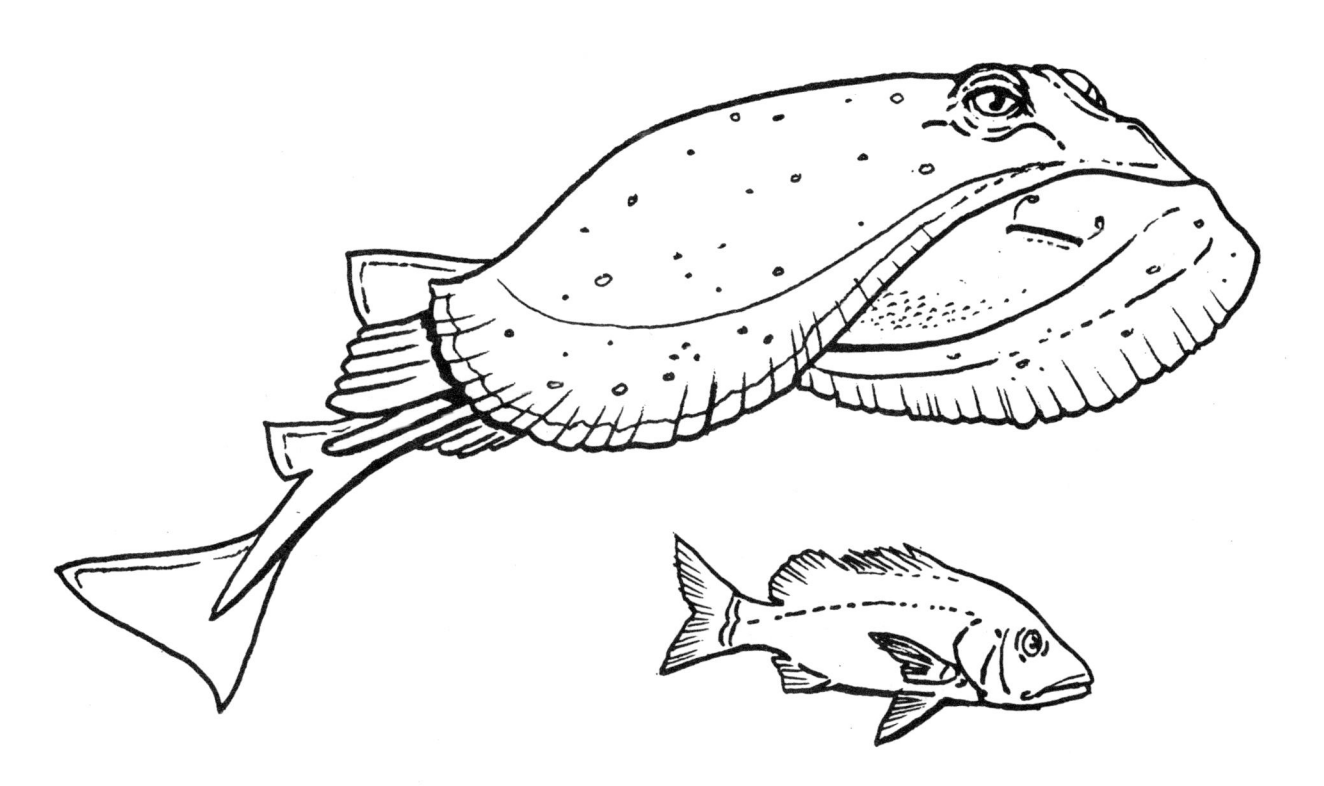

LITTLE SKATE
Raja erinacea

Little skates grow to be 21 inches long and only weigh one pound. It has brown spots on its back and wings. Rows of pointed thorny spines grow along its back and down its tail. The little skate lives in shallow water feeding on clams, squid, worms and crustaceans.

••

The little skate is found along the North American Atlantic Coastline.

LITTLE SKATE
Raja erinacea

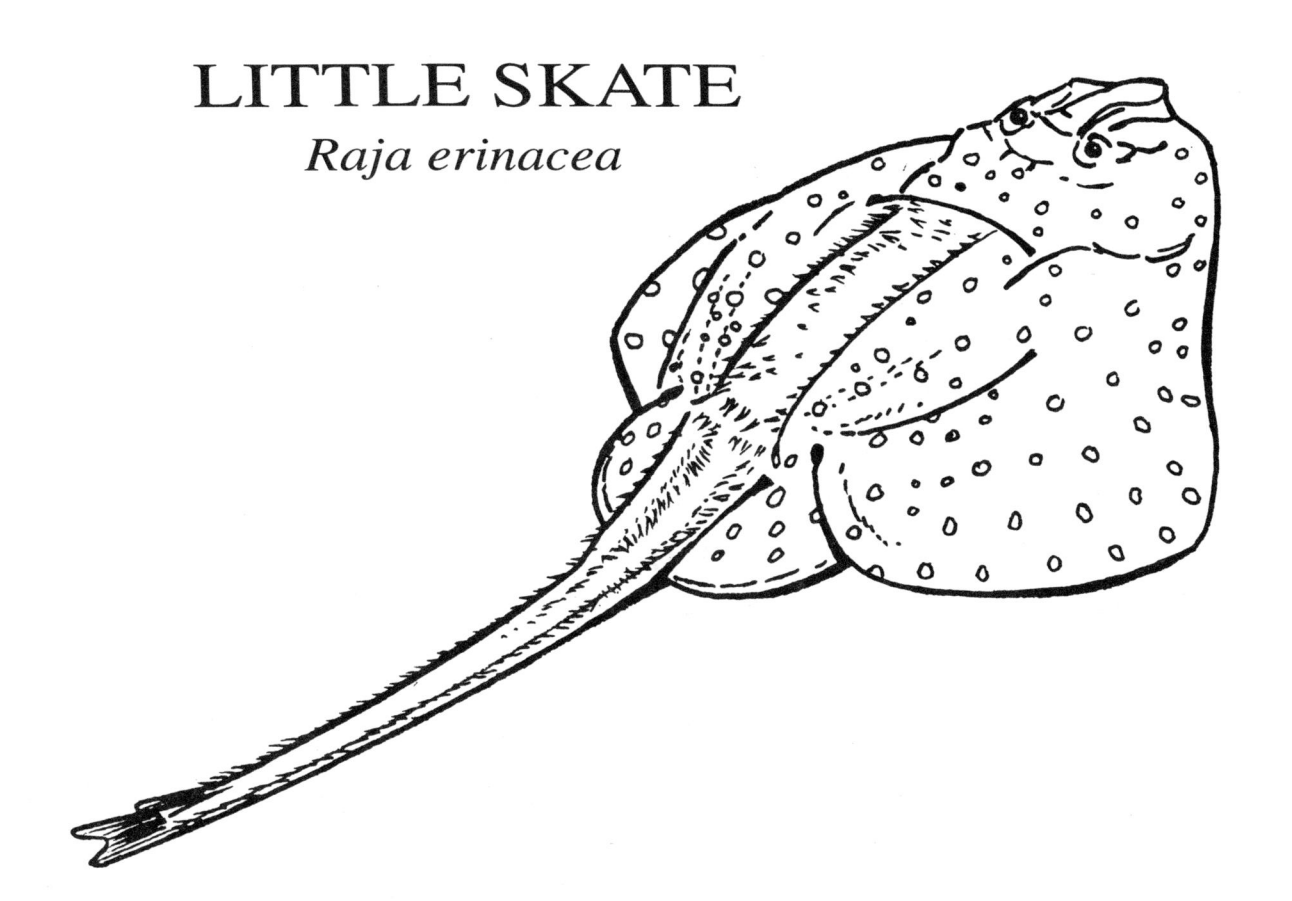

SPOTTED EAGLE RAY
Aetobatus narinari

Spotted eagle rays prefer to be alone but will gather into schools during spawning or for migration. A series of short poisonous spines at the base of its tail are used for protection. These graceful rays ride the oceans currents, like birds in the air do. Occasionally they break the waters surface and "glide" through the air.

....

This ray is found in the Gulf of Mexico, the Bahamas and the Atlantic Ocean from Massachusetts to Brazil.

SPOTTED EAGLE RAY
Aetobatus narinari

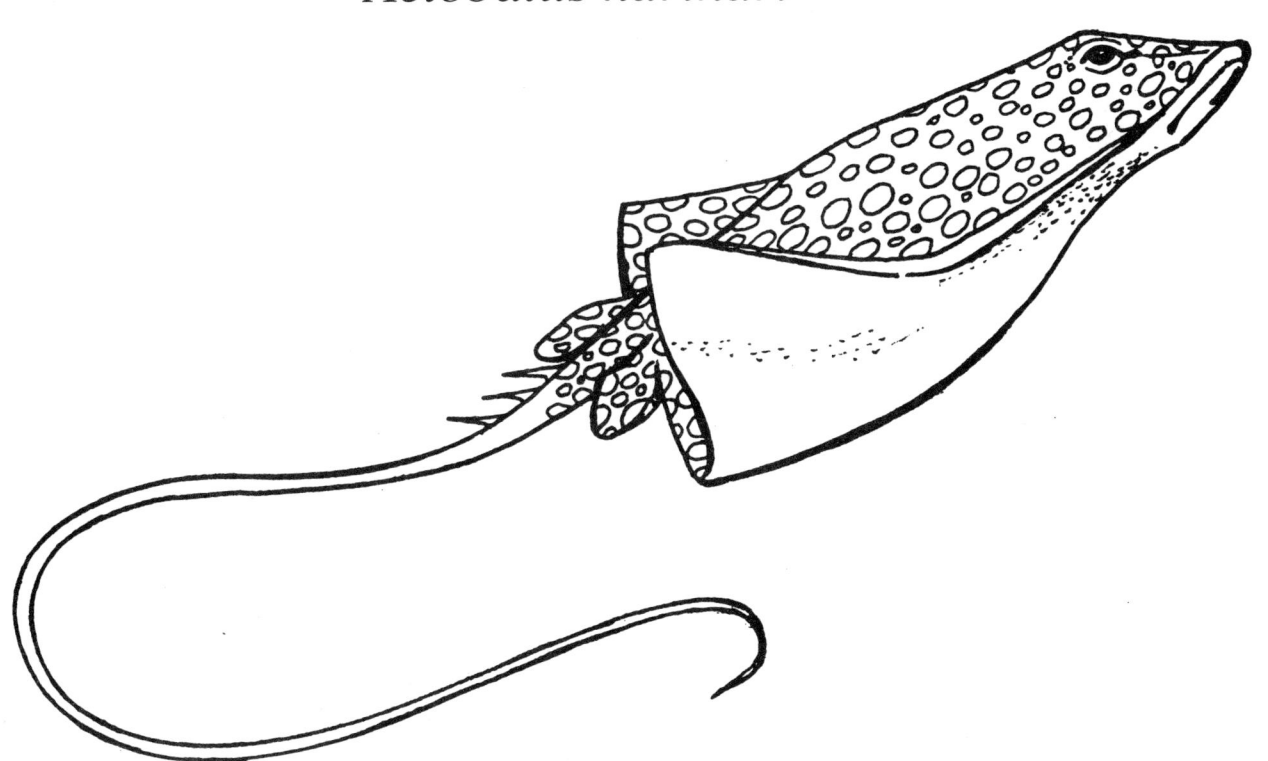

ATLANTIC COWNOSE RAY
Rhinoptera bonsus

The most common of the eagle rays, the cownose ray can have a wingspan of 6 feet. Behind its dorsal fin is a poisonous stinging spine. This stinging spine makes the cownose dangerous to swimmers and waders. It swims in shallow waters looking for clams and oysters, which it crushes in its toothplates.

••

The Atlantic cownose ray is found in the Atlantic Ocean from New England to Brazil and the Gulf of Mexico.

ATLANTIC COWNOSE RAY
Rhinoptera bonsus

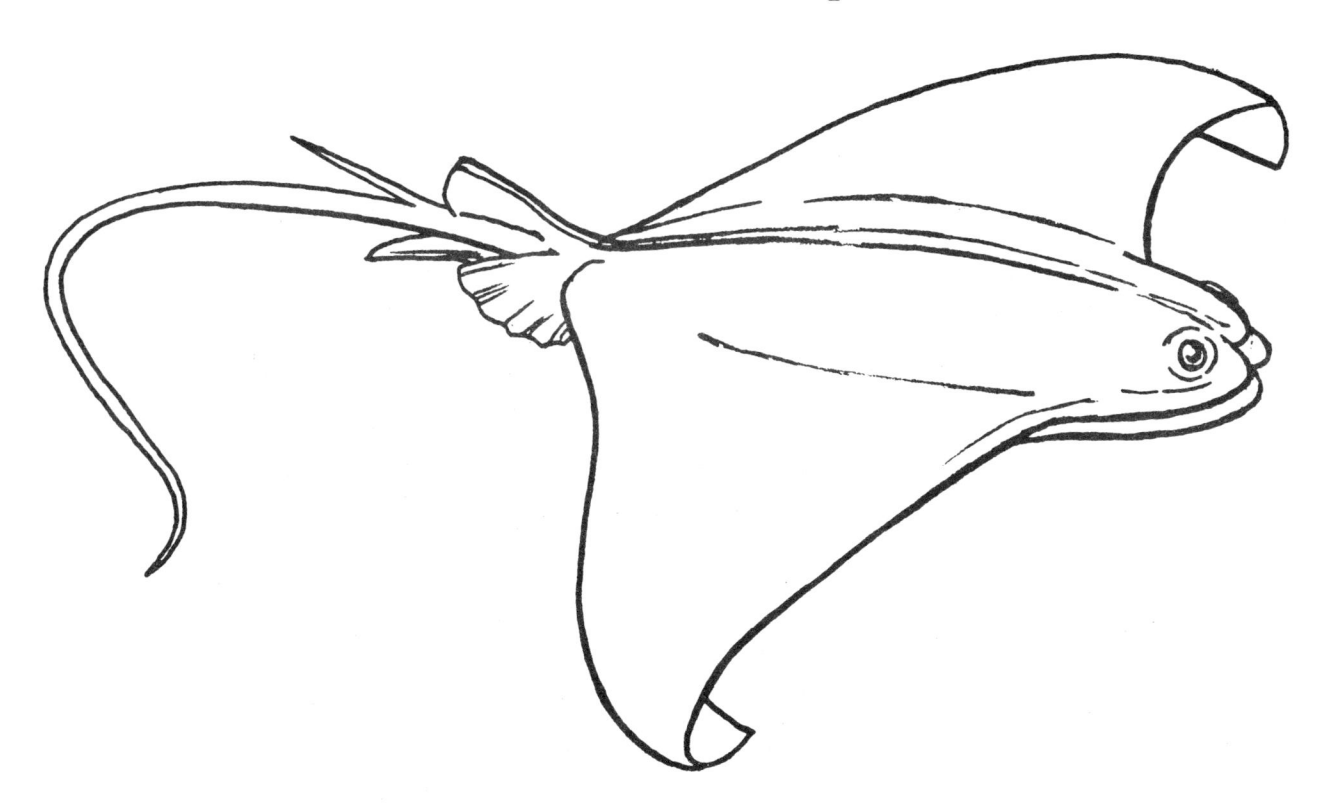

ATLANTIC MANTA
Manta birostris

This giant manta has a wing span of 22 feet and can weigh up to 300 pounds. Even though it lives in deep water, it can often be seen sunning itself on the surface, with its "wings" outspread. Don't let its large size scare you - it does not seem harmful to man. It survives on plankton, small fish and crustaceans.

•••

The Atlantic Manta is found from New England to Brazil in the Atlantic Ocean and in the Gulf of Mexico.

ATLANTIC MANTA
Manta birostris

COMMON SKATE
Raja batis

This 18 inch skate prefers the shallow waters close to Europe's shores. As bottom dwellers, they lay half buried on the bottom during the day. At night they become active to feed on their favorite crustaceans and shell fish. The common skate is recognized by its large wing-like pectoral (chest) fins.

• •

The common skate is found off the coast of Europe and is harvested commercially.

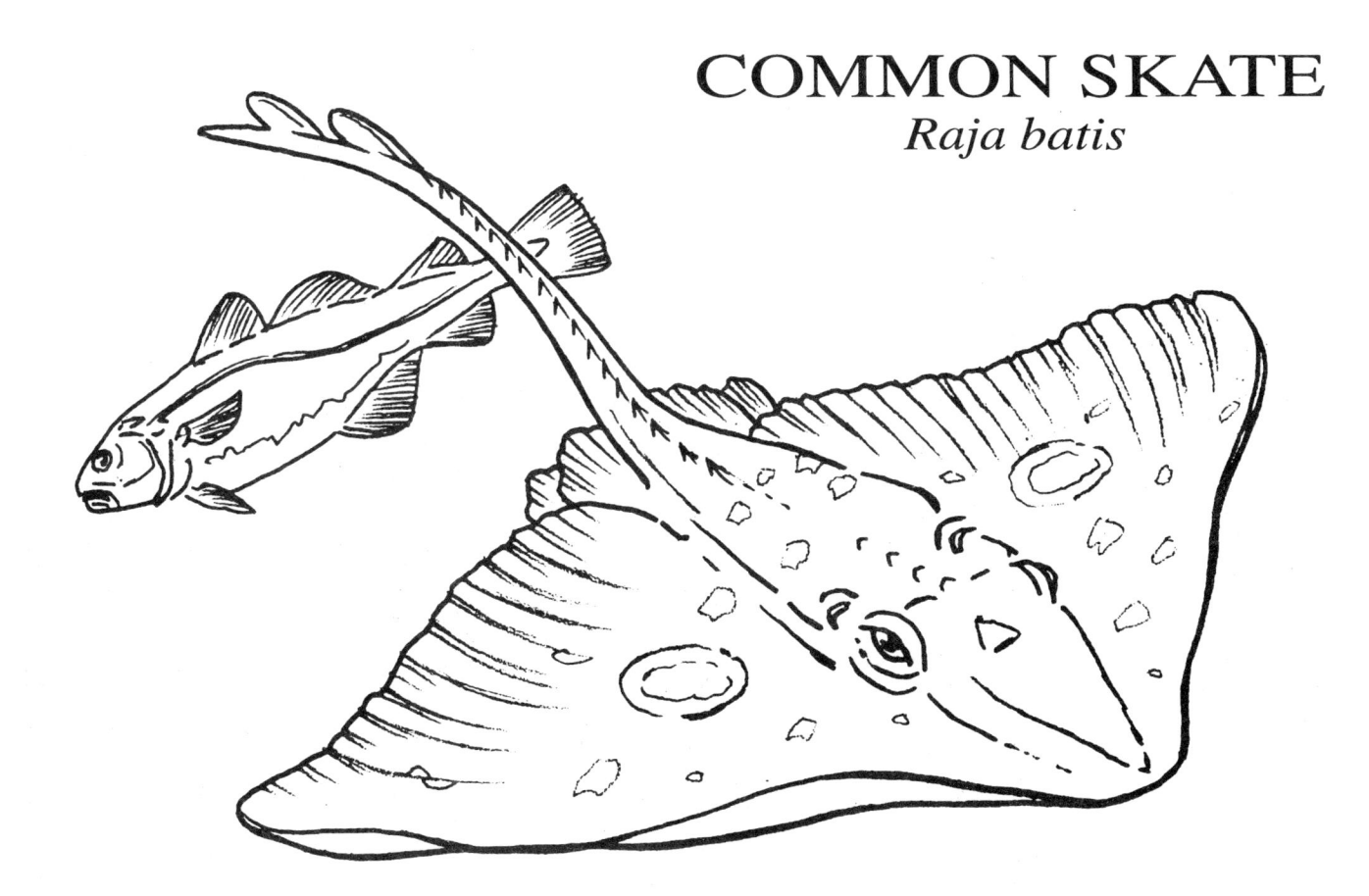

COMMON SKATE
Raja batis

OTHER CHILDREN'S BOOKS CREATED BY SPIZZIRRI PUBLISHING

ISBN (INTERNATIONAL STANDARD BOOK NUMBER) PREFIX ON ALL SPIZZIRRI BOOKS IS: 0-86545-

EDUCATIONAL READ AND COLOR BOOKS
ILLUSTRATIONS AND TEXT
SIZE: 8 1/2" X 11"

- AIRCRAFT
- ANIMAL ALPHABET
- ANIMAL F. CALENDAR
- ANIMAL GIANTS
- ATLANTIC FISH
- AUTOMOBILES
- BIRDS
- CALIFORNIA INDIANS
- CALIFORNIA MISSIONS
- CATS
- CATS OF THE WILD
- CAVE MAN
- COLONIES
- COMETS
- Count/Color DINOSAURS
- COWBOYS
- DEEP-SEA FISH
- DINOSAURS
- DINOSAURS OF PREY
- DOGS
- DOGS OF THE WILD
- DOLLS
- DOLPHINS
- EAGLES
- ENDANGERED BIRDS
- Endang'd Mam'ls-AFRICA
- Endang'd Mam'ls- ASIA & CHINA
- Endang'd Mam'ls-SO. AMERICA
- ENDANGERED SPECIES
- ESKIMOS
- FARM ANIMALS
- FISH
- HORSES
- KACHINA DOLLS
- LAUTREC POSTERS
- MAMMALS
- MARINE MAMMALS
- MARSUPIALS
- NORTHEAST INDIANS
- NORTHWEST INDIANS
- PACIFIC FISH
- PALEOZOIC LIFE
- PENGUINS
- PICTURE CROSSWORDS
- PICTURE DICTIONARY
- PIONEERS
- PLAINS INDIANS
- PLANETS
- POISONOUS SNAKES
- Prehist. BIRDS
- Prehist. FISH
- Prehist. MAMMALS
- Prehist. SEA LIFE
- PRIMATES
- RAIN FOREST BIRDS
- RAIN FOREST RIVER LIFE
- RAIN FOREST TREE LIFE
- REPTILES
- ROCKETS
- SATELLITES
- SHARKS
- SHIPS
- SHUTTLE CRAFT
- SOUTHEAST INDIANS
- SOUTHWEST INDIANS
- SPACE CRAFT
- SPACE EXPLORERS
- STATE BIRDS
- STATE FLOWERS
- TEXAS
- TRANSPORTATION
- TRUCKS
- WHALES

SILHOUETTE ART BOOKS
8 1/2 x 11" Reproducible

- CHRISTMAS
- CIRCUS
- DINOSAURS
- FARM ANIMALS
- OCEAN LIFE
- ZOO ANIMALS

EDUCATIONAL ACTIVITY BOOKS
SIZE: 5 1/2 x 8 1/2"

- Alphabet Dot-to-dot PETS
- Alphabet Dot-to-dot ZOO ANIMALS
- BIRD MAZES
- BUTTERFLY MAZES
- DINOSAUR MAZES
- Dot-to-dot DINOSAURS
- Dot-to-dot FISH
- Dot-to-dot REPTILES
- Dot-to-dot WHALES
- FARM MAZES
- FISH MAZES
- FLOWER MAZES
- MAMMAL MAZES
- SHARK MAZES
- SHELL MAZES
- TREE MAZES
- TURTLE MAZES
- ZOO MAZES

Educational FACTS and FUN Books
SIZE: 5 1/2 X 8 1/2"

- ANIMAL LEAPS
- ANIMAL SPEEDS
- ANIMALS THAT LAY EGGS
- ANIMALS WITH LONG NECKS
- ANIMALS WITH LONG TAILS
- BIRD SPEEDS

EARLY LEARNING WORKBOOKS
SIZE: 8 1/2 x 11"

- ALPHABET PICTURES
- COMPLETE THE WORDS
- COUNTING DINOSAURS
- Dot to dot ALPHABET
- Dot to dot NUMBERS
- FIRST ADDITION
- FIRST ALPHABET
- FIRST NUMBERS
- FIRST SUBTRACTION
- MAKE A CALENDAR
- MAKING WORDS
- MAZE PUZZLES
- NUMBERS and COLORS
- PICTURE CROSSWORDS
- PICTURE DICTIONARY
- SHAPES, ART & COLORS
- THEY GO TOGETHER
- TRIANGLE PICTURES
- WORD HUNT PUZZLES
- WORDS IN WORDS

- FOOTPRINTS OF BIRDS
- INSECT HUNTERS
- MAMMAL FOOTPRINTS
- POISONOUS ANIMALS
- SKATES AND RAYS
- SMALL MAMMALS